अंकांची गोष्ट

THE NUMBER STORY

SMALL BOOK ONE

ENGLISH - MARATHI

---

*Numbers Teach Children
Their Number Names*

written and illustrated by

MISS ANNA

Early Reader Edition of *The Number Story 1*
Bronze Medal Winner, 2016 Wishing Shelf Book Award

Library of Congress Control Number: 2018902040

Names: Miss Anna, author.
Title: Number story : numbers teach children their number names / Miss Anna.
Description: Portland, OR: Lumpy Publishing, 2018.
Identifiers: ISBN 978-1-945977-90-9| LCCN 2018902040
Summary: The pictures and rhymes present stories which introduce numbers 0-10.
Subjects: LCSH Numeration—English--Marathi--Pictorial works--Juvenile literature. | BISAC JUVENILE NONFICTION /
Languages: English--Marathi
Classification: LCC QA141.3 .M57 2018 | DDC 513—dc23

Publisher: Lumpy Publishing
Website: www.missannabooks.com
Email: missanna@missannabooks.com

Paperback: ISBN 978-1-945977-90-9
Printed in the U.S.A.   1 3 5 7 9 10 8 6 4 2

Want to learn our number names?

तुम्हाला अंकांची नावे शिकायला आवडेल का?

It is very easy and a lot of fun!

हे तर खूपच सोपे आणि मजेशीर आहे.

Say-along our little jingle

म्हणा तर मग आमच्यासोबत ह्या
छोट्या गोष्टीचे गाणे!

starting from Number One!

आपण एक या अंकापासून सुरुवात करूया!

1

ONE   looks like my one finger.

१ ☆ एक

दिसतो माझ्या एका बोटासारखा

ONE!
एक!

2

२ ✧ दोन

शेपटीचा माग काढा

A TAIL! शेपूट!

THREE   has bumps.

३ ☆ तीन

याला आहेत चढ उतार

चढ उतार!

4

FOUR   carries a sail.

४ ☆ चार

एक जहाज चालते

4
A SAIL!
जहाज!

# 5

FIVE   is a racing track.

५ ☆ पाच

हा तर धावण्याचा ट्रॅक

VROOM!
THHHH!
1

6

SIX   curves like a snail.

६ ☆ सहा

गोगलगायीसारखा वळणदार

A SNAIL! गोगलगाय!

7

७ ★ सात

याचे कोन टोकदार

BE CAREFUL! IT'S SHARP!

काळजी घ्या! हा आहे टोकदार!

8

EIGHT   is rollercoaster rails.

८ ✦ आठ

उलटीपालटी गुंडाळी आगगाडी

येह!

यीप्पी!

YIPPEE!

9

NINE   is a bubble on a stick.

९ ☆ नऊ

काठीवरचा बुडबुडा

A BUBBLE! बुडबुडा!

10
TEN   is an eye of a whale.
१० ☆ दहा
व्हेल चा एक डोळा

मिचकवा!
WINK!

HELLO! हॅलो!

And
आणि
0
ZERO   is an empty pail.
० ☆ शून्य
एक रिकामा घडा

IT'S EMPTY!
हा तर रिकामा!

Thank you for playing with us today.

We had a lot of fun too!

धन्यवाद, आमच्यासोबत आज खेळल्याबद्दल.

आम्हालाही खूप मज्जा आली!

We are your Number friends,
Zero to Ten,
Who will be here for you~
आम्ही आहोत तुमचे अंक मित्र
शून्य ते दहा.
आम्ही इथे तुमच्यासाठी नेहमीच असू.

Bye-bye now!
See you again soon.
आत्ता अच्छा, टाटा, बाय!
परत लवकरच भेटू!

The Numbers are *SINGING* too!

To sing-a-long, look for Miss Anna Number Story
at your favorite music store like iTUNES.

MP3

Numbers 0-10
IDENTIFYING
& COUNTING

Numbers 11-20
& Ordinals
first, second, third...

Numbers 0-100
& Place Values
ones, tens, hundreds...

About Clocks
& Telling Time
hours, minutes, seconds

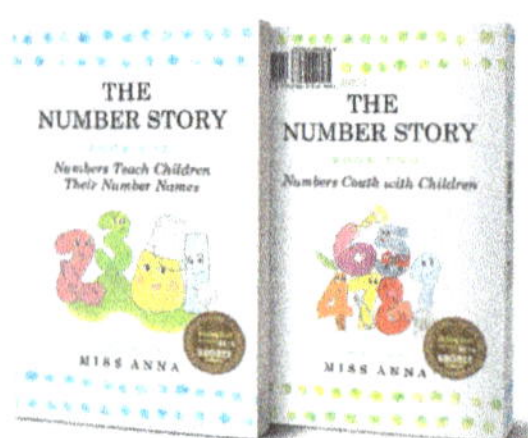

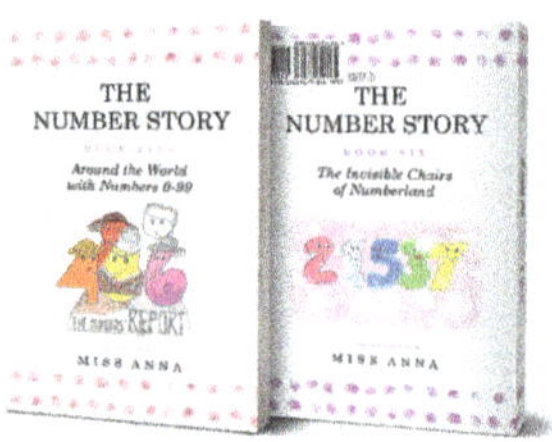

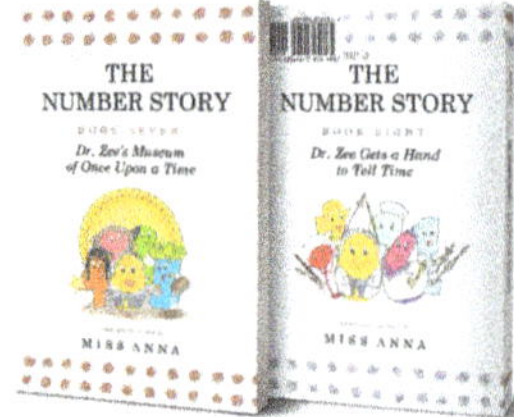

Number Story 1 & 2
isbn: 978-0-996216-48-7

Number Story 3 & 4
isbn: 978-1-945977-01-5

Number Story 5 & 6
isbn: 978-1-945977-06-0

Number Story 7 & 8
isbn: 978-1-949320-40-4

For more Miss Anna books to love,
visit us at

www.missannabooks.com

Numbers are working hard all over the world!
Come Travel the World with Us!

www.ingramcontent.com/pod-product-compliance
Lightning Source LLC
Chambersburg PA
CBHW040901070726
47599CB00035B/2258